My Uniform Pocket

is dedicated to all of the awesome
military kids out there (and their dads).

Find more, including additional note and
drawing pages to fill **your** uniform pocket as well
as flight suit and mom editions of this book, at
BrainExecutiveProgram.com/MyPocket

My dad is in the military.

That means he works to help keep our country and the whole world safe.

safer because of
my dad

He has a really special job,
so he wears a uniform.

dad's
uniform

my
uniform

His uniform helps him match everyone at
work, so they look like a team.

The military has many kinds of uniforms. Some are for everyday work and some are for dressing up,

but they all have one thing in common...

lots of pockets!

Can you find the pockets on your dad's uniform?

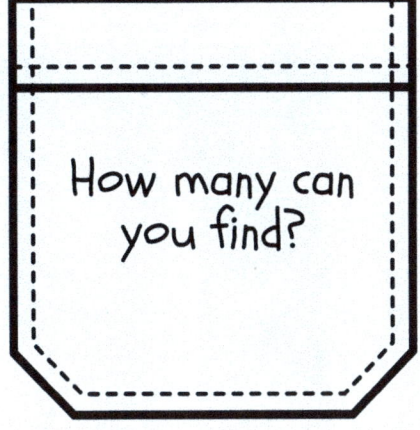

How many can you find?

My dad fills his uniform pockets
with lots of stuff.

This uniform has a
small pocket for
pens and pencils
on the sleeve.

Some uniforms have hidden pockets where he can carry his wallet.

Where does your dad keep his wallet?

Other uniforms have a pocket where he puts a special hat called a cover,

and he always finds a pocket
where he can keep snacks.

No matter what uniform my dad wears,
he has one pocket that is really special.

It's always
the one closest
to his♡.

He gave it to me. It's **my** uniform pocket.

Can you find the pocket closest to the heart for each uniform?

Now, find it on your
dad's uniform!

Sometimes my dad has to be gone for a really long time because of his job.

When that happens, I miss him a lot.

I wish I could go with him...

but he says that kids are not allowed.

That is why he gave one of his
uniform pockets to me.

It's a special pocket that I put
things in just for him.

Sometimes I give my dad notes
and cards to put in my pocket.

Sometimes I make him
drawings or give him photos.

Sometimes I even find small things to send
him like a tiny rock I found at the park
or a shell from the beach.

It makes me feel
better that even when
my dad is far away,

he has something with him
that is from me.

He says that it makes us feel close...

even when we are far apart.

Having a dad in the military
is really special.

My dad says that I help keep the
whole world safe

by letting him go to work.

And while we are far apart, I know that he has something from me close to his heart,

in **my** uniform pocket.

Dear Daddy,

Me and Dad

Dear Daddy,

My Family

Hi Dad,

Also by Kathryn Hamlin-Pacheco

The same endearing story as found in *My Uniform Pocket* but written for families who wear flight suits.

A workbook for kids to help them understand their sensory processing patterns.

Book Cover by Kathryn Hamlin-Pacheco
Illustrations by Kathryn Hamlin-Pacheco

www.ingramcontent.com/pod-product-compliance
Lightning Source LLC
Chambersburg PA
CBHW051338120626
46547CB00016B/2594